21ST Century Skills **INNOVATION LIBRARY** Disruptors in Tech

3D Printing

Martin Gitlin

Published in the United States of America by Cherry Lake Publishing
Ann Arbor, Michigan
www.cherrylakepublishing.com

Reading Adviser: Marla Conn, MS, Ed., Literacy specialist, Read-Ability, Inc.

Photo Credits: ©YAKOBCHUK VIACHESLAV/Shutterstock, cover, 1; ©bissig/Shutterstock, 5; ©Historic American Engineering Record/Library of Congress/Control No. oh1541, 6, 7; ©The Advertising Archives/Alamy Stock Photo, 8; ©Marina_Skoropadskaya/iStock, 10; ©Tinxi/Shutterstock, 13; ©Cholpan/Shutterstock, 14; ©MarinaGrigorivna/Shutterstock, 15, 25; ©Scharfsinn/Shutterstock, 19; ©Steven Keating, 2016/DCP printing upper layers of dome, The Mediated Matter Group, 20; ©NASA, 21; ©Volodymyr Nik/Shutterstock, 22; ©Steve Heap/Shutterstock, 26

Graphic Element Credits: ©Ohn Mar/Shutterstock.com, back cover, multiple interior pages; ©Dmitrieva Katerina/Shutterstock.com, back cover, multiple interior pages; ©advent/Shutterstock.com, back cover, front cover, multiple interior pages; ©Visual Generation/Shutterstock.com, multiple interior pages; ©anfisa focusova/Shutterstock.com, front cover, multiple interior pages; ©Babich Alexander/Shutterstock.com, back cover, front cover, multiple interior pages

Copyright © 2020 by Cherry Lake Publishing

All rights reserved. No part of this book may be reproduced or utilized in any form or by any means without written permission from the publisher.

Library of Congress Cataloging-in-Publication Data

Names: Gitlin, Martin, author.
Title: 3D printing / by Martin Gitlin.
Other titles: 3-D printing
Description: Ann Arbor : Cherry Lake Publishing, [2019] | Series: Disruptors in tech | Audience: Grades: 4 to 6. | Includes bibliographical references and index.
Identifiers: LCCN 2019006014 | ISBN 9781534147621 (hardcover) | ISBN 9781534150485 (pbk.) | ISBN 9781534149052 (pdf) | ISBN 9781534151918 (hosted ebook)
Subjects: LCSH: Three-dimensional printing—Juvenile literature.
Classification: LCC TS171.95 .G58 2019 | DDC 621.9/88—dc23
LC record available at https://lccn.loc.gov/2019006014

Printed in the United States of America
Corporate Graphics

Martin Gitlin has written more than 150 educational books. He also won more than 45 awards during his 11-year career as a newspaper journalist. Gitlin lives in Cleveland, Ohio.

Table of Contents

Chapter One
Before the Breakthrough 4

Chapter Two
How 3D Printing Works 12

Chapter Three
Endless Potential 18

Chapter Four
A Look to the Future 24

Timeline 28
Learn More 30
Glossary 31
Index 32

CHAPTER ONE

Before the Breakthrough

Printers and copiers have made life easier. If you leave your assignment at home, you can ask someone to scan and email it to you, and you can print it at school. If you need to make 40 copies of a document, you can do that with just a few clicks of a button. Who invented the printer and copier that have made life easier? What new advancements are there to these machines?

The Start of Something New

It was 1937 and Chester Carlson was frustrated with his job. He worked in New York City. He had to make copies of many important papers. The task was hard and expensive. Carlson decided to take action. After all, he was an inventor.

Carlson experimented with **static electricity** to transfer images from one piece of paper to another. One attempt filled his apartment with the stinky smell of sulfur and forced him to move. Carlson kept trying.

Chester Carlson

USA 21

Chester Carlson used the static electricity from a handkerchief, light, and special powder to make the world's first photocopy!

Although, Carlson created the first photocopy in 1938, the copier itself, called the Xerox machine, wasn't released to the public until 1959!

He finally created the first copier through a process called xerography. Carlson made the first photocopy in history on October 22, 1938. The copier became one of the most well-known inventions of the century and made Carlson a very rich man.

But that wealth did not happen immediately. People did not grasp the value of the copier at first. It took Carlson 4 years to find an institute to develop a copy machine. Then, in 1947, the Haloid Company in New York city approached Carlson. It wanted to help bring Carlson's copier to offices everywhere. That business later changed its name to the now-famous Xerox Corporation.

Xerox released the world's first color copier in 1973.

The first commercially available Xerox machine weighed almost 650 pounds (294.8 kilograms).

Printing Quickly

Carlson had begun to make life easier for workers around the world. His invention led to the first high-speed printer in 1953. The first push-button plain paper copier hit the market in 1959. It sold like crazy. The income of the Haloid Company soared from $2 million in 1960 to more than $22 million in 1963.

New ways to print and copy began hitting the market. The first **laser** printer was released in 1976. Among the most successful companies was Canon, which became the leading producer of copiers by 1985. At that time, these copiers and printers could only be found at businesses. It wasn't long, though, before they were introduced for personal use in homes.

During this time, another type of printer was introduced. In 1986, Chuck Hull created the first actual 3D printer and revolutionized the world.

> Sometimes a company becomes so famous that its name is also referred to as the product it makes. One example is Kleenex. Another is Xerox. Many people still call a photocopy a "xeroxed" copy, even if it was not produced on a copier made by Xerox.

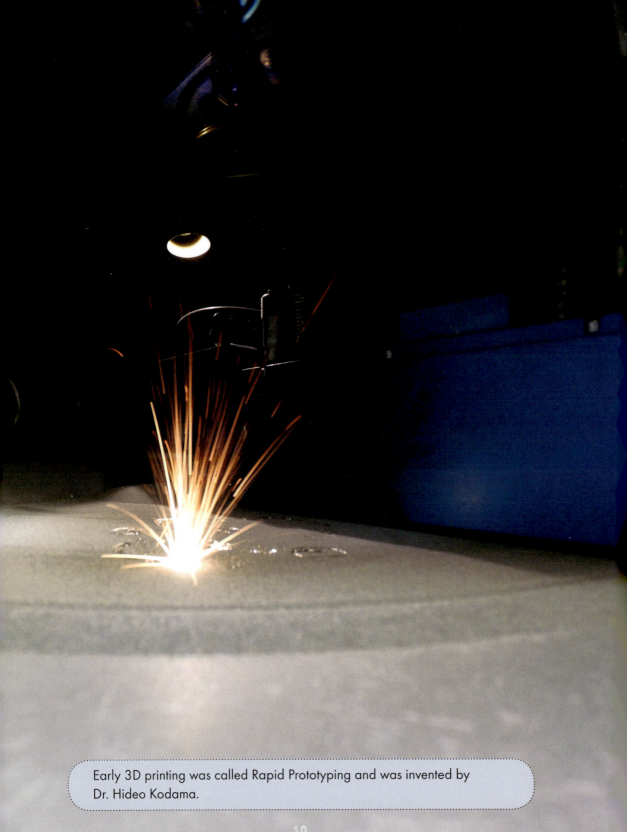

Early 3D printing was called Rapid Prototyping and was invented by Dr. Hideo Kodama.

The Lowdown on Laser Printing

The biggest scientific news of 1969 was never going to be the first laser printer design. That's because this was also the year of the first manned moon landing. But Gary Starkweather did make a splash.

Starkweather developed the first laser printer while working at the Xerox research facility. He gained credit for using a laser to re-create an image on a copier drum to print onto paper. The laser printer was initially **manufactured** in California. But it was IBM that first placed one on the market in 1976. Xerox followed suit a year later.

The invention could print out copies much faster than previous models. The computer revolution brought more companies into the laser printing business. Hewlett-Packard and Apple both produced them in 1985. The Hewlett-Packard Laser Jet was the most famous, becoming the first LaserJet printer to be significantly used in homes.

CHAPTER TWO

How 3D Printing Works

Want to print a human ear? A necklace? Chocolate cake? Parts for a spaceship? Even in the late 20th century, no one would have imagined this is where we'd be with 3D printing **technology**. Yet it is becoming a reality.

Traditional vs. 3D Printers

Traditional printers do not show images in 3D. That stands for three-dimensional, which gives an image or object height, width, and depth. A 3D printer creates objects one layer at a time like a kid building with Legos.

Different professionals work in 3D. In the past, architects had to build their models of homes by hand. When 3D printers were invented, they could use them to build their models 10 times faster and at far less cost. Car companies, like Volvo and Mercedes-Benz, also rely on 3D printers to reduce time and cut back on cost.

MakerBot, a 3D printing company, was one of the first in the industry to share the technology with the general public.

Some 3D printers use recycled plastic!

No Ink Necessary

Unlike traditional printers, 3D printers do not use ink. A 3D printer deposits layers of material that melt when heated through a tiny nozzle. The layers are then fused together. The material used in 3D printers is solid when at room temperature. When used inside the printer, the material melts at 220 degrees Fahrenheit (104 degrees Celsius). Once it is extracted from the printer, it can be sanded smooth or painted any color.

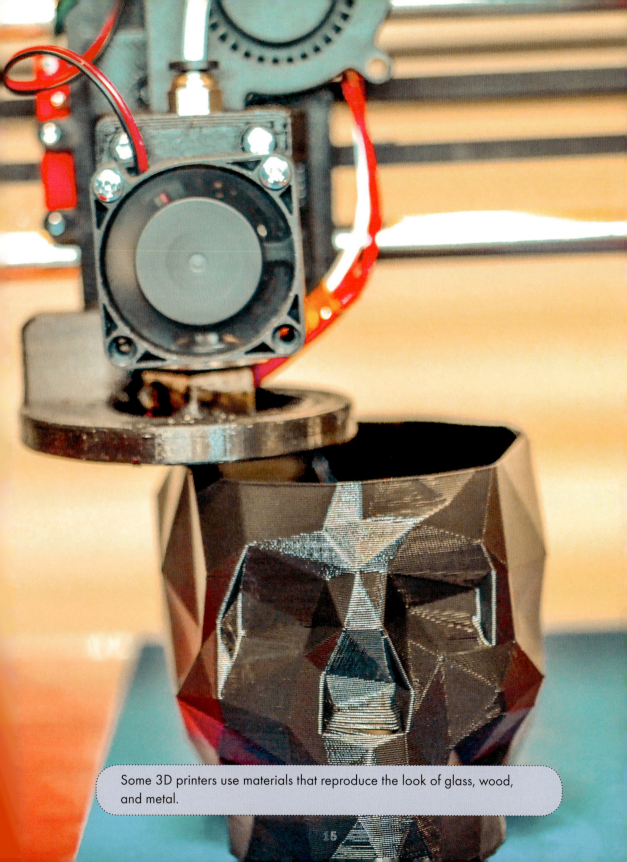

Some 3D printers use materials that reproduce the look of glass, wood, and metal.

Advantages and Disadvantages

One of the many advantages of a 3D printer is its speed. It can produce a product in hours rather than days. The printers are expensive, but they're much cheaper than machines that manufacture material. According to sources, industrial 3D printers, like those used by big companies, can cost between $20,000 to $200,000. But other types of machines that make similar products might cost 10 times as much!

The models produced by 3D printers are usually inferior to those made in manufacturing plants. But the technology of 3D printing is new, and innovators have barely touched the surface of what the printers can create. What is already known, however, is amazing. These marvels have made an impact on many industries.

> Online stores such as Etsy have provided an outlet for people to make and sell crafts from home. New technology has created the same options for those great at 3D printings. Online sites such as Zazzle and Shapeways give people opportunities to make and sell their own 3D-created arts and crafts.

Charles Hull

In the mid-1980s, Charles Hull developed the solid imaging process that made 3D printing technology possible. Hull designed a technique for creating 3D objects controlled by a computer. The objects were built layer by layer using a liquid that hardened when contacted by laser light. Printers that used this technology were expensive. But Hull felt strongly enough about its potential to launch a company called 3D Systems in 1986. That business kicked off the 3D printing industry.

Hull gained fame with his invention. But he did not stop there. He was named inventor on more than 60 American patents and others around the world. It is no wonder that he was inducted into the National Inventors Hall of Fame in 2014.

CHAPTER THREE

Endless Potential

The number of ways that 3D printing can help people is endless. The technology promises to aid many industries. Let's take a look at a few.

Health Care

The most important industry 3D printing impacts is health care. Doctors have been among the first to explore its uses. A 3D printer has produced drugs that can improve health. It has also printed ears and boasts the potential to produce other body parts such as arms and legs for those who have lost theirs in an accident. The technology of 3D printing allows those body parts to fit people of different shapes and sizes perfectly.

Organs such as hearts and lungs cannot yet be created through 3D printing. But a project at Wake Forest University in North Carolina called Body on a Chip is providing hope. It has printed tiny human hearts, lungs, and blood vessels. These were placed on **microchips** and tested with **artificial** blood. Experiments with brain surgery have also utilized 3D printing.

3D printing in medicine is called "bioprinting."

Building Buildings

The invention of the 3D printer helps charitable organizations and causes. 3D printers can build shelters quickly and cheaply for victims of hurricanes or other natural disasters. They can also build places for homeless people to live.

> It might be a while before 3D printing is ready to produce body parts such as hearts and lungs. But 3D printers have been used for medical education and training. Doctors at one hospital in Florida have practiced surgical techniques on 3D-printed replicas of children's hearts. Other hospitals have used 3D printers to practice brain surgery.

In 2017, researchers from the Massachusetts Institute of Technology invented a 3D printing robot that built a 50 feet (15 meters) wide and 12 feet (4 m) high building in less than 14 hours!

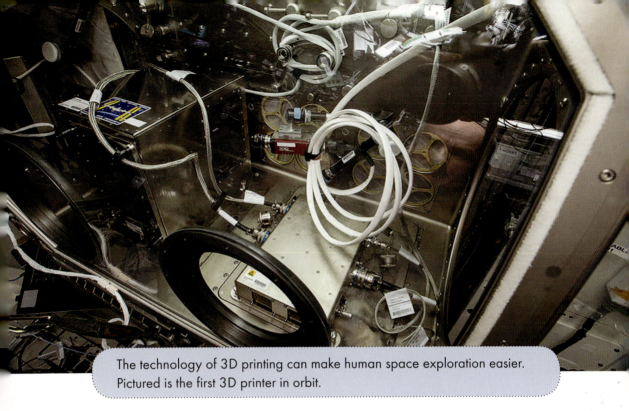

The technology of 3D printing can make human space exploration easier. Pictured is the first 3D printer in orbit.

Air and Space Travel

The world of air and space travel seems certain to benefit from the new technology. Some large airplanes can have more than 300,000 parts. 3D printing makes it possible to design, test, and manufacture those parts quickly and cheaply.

It will also allow parts of a spaceship to be created outside of Earth. Today, an astronaut who is millions of miles away in space and looking at a broken part could be doomed. But in the future, that space traveler will be able to use a 3D printer to produce an identical and perfect part.

Some artists may choose to 3D print their creation using cement.

Calling All Artists

Not all uses of 3D printing will be practical. Some will help make the world a more beautiful place. The new technology can help artists create paintings or sculptures. It could also guard against the creation of faked copies.

A future generation of artists might create works very differently. Painters might scrap the canvas. Sculptors might put down the chisel. Both might instead use a 3D printer to produce art layer by layer.

Using 3D printing could also allow two artists who live thousands of miles apart to work on the same project. They would simply share their files and print out identical work.

CHAPTER FOUR

A Look to the Future

There's hardly an industry that wouldn't be impacted by 3D printing. A toy builder could create a new doll. A furniture maker could manufacture a new couch. A fashion designer could make a beautiful dress. A sporting goods store could produce its own baseball bats. All of this can be achieved through 3D printing.

Bringing It Home

But its potential does not end in the business world. People can use 3D printers around the home. If they break a vase, they can design and build a new one. If they need a new chair, a 3D printer can build one for them. If they outgrow their shoes, they can manufacture a pair exactly one size larger.

The world of 3D printing could even change the world of food. Can a person eventually make all their own food rather than shop at the grocery store? What will become of supermarkets? And will there be any need for farmers to grow food?

Some companies have invented 3D printers that work with fresh food and ingredients. Pictured is a 3D printer printing pancakes.

One curious journalist had fun with 3D printing. *The New York Times* writer A. J. Jacobs tried to print out an entire meal, which included pizza and an eggplant dish, in 2013. He even printed out the plates and silverware! The dinner was a success. But Jacobs admitted that it required a lot more work than preparing a meal the usual way.

There are 3D printing devices, like 3D pens, that are for people of all ages.

Still in Business

You might wonder what will happen to businesses that make and sell goods that people will be able to create at home. In reality, these businesses do not have much to worry about. Time and cost will limit what people can produce at home. The 3D printer will benefit businesses the most—they will purchase the printers to make things they can sell.

However, most businesses are not yet ready to use 3D printing. Buying 3D printers is cheaper than setting up a factory. But the cost is still far more than using a traditional method of production. And the quality of 3D printer products cannot yet compare.

The future of 3D printing is promising. Imagine a world with nothing ever out of stock. Imagine a world in which you break a window and manufacture a new one yourself. Even better, imagine a world with far less waste, packaging, and pollution.

How About a Printed Dinner?

The potential impact of 3D printing on food worldwide is staggering. One can imagine it ending world hunger. Printers could someday cook and serve food on a massive scale. It is believed they could improve the **nutritional** value of meals to make people healthier. They could feed people in areas of the world that lack access to fresh and affordable ingredients.

Modern 3D food printers feature nozzles, powders, lasers, and robotic arms. They can create patterned chocolate and pastries. One food printer uses ingredients loaded onto steel capsules to make foods such as pizza, pasta, and brownies. Others print nutrition bars. Restaurants, bakeries, and candy makers have already used 3D printers to save time and energy.

The world's population is expected to soar to nearly 10 billion by 2050. Food production must keep up with demand. That will make 3D printers a necessity. Not only can they create food for individual tastes and nutritional needs, they can also mass-produce enough food to feed everyone.

Timeline

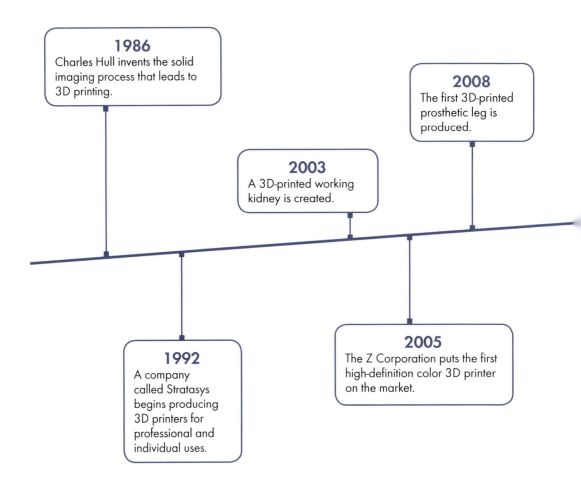

1986
Charles Hull invents the solid imaging process that leads to 3D printing.

2003
A 3D-printed working kidney is created.

2008
The first 3D-printed prosthetic leg is produced.

1992
A company called Stratasys begins producing 3D printers for professional and individual uses.

2005
The Z Corporation puts the first high-definition color 3D printer on the market.

2010
Urbee becomes the first 3D-printed prototype car.

2011
Cornell University debuts the first 3D food printer.

2012
The first prosthetic jaw is printed and implanted.

2013
President Obama expresses hope for the future by mentioning 3D printing in his State of the Union address.

2015
Carbon 3D releases its super-fast CLIP 3D printer.

Learn More

Books

Abell, Tracy. *All About 3D Printing.* Mendota Heights, MN: Focus Readers, 2017.

3D Printing Projects. New York, NY: Dorling Kindersley, 2017.

Websites

All3DP
https://all3dp.com/3d-printing-with-kids-what-you-need-to-know
This site provides readers a variety of information about 3D printing.

Beanz
https://www.kidscodecs.com/resources/3d-printing
This online computer science magazine provides links that explain much about 3D printing.

Glossary

artificial (ahr-tuh-FISH-uhl) made by human beings often following a natural model or process

laser (LAY-zur) a device that produces a very narrow, intense beam of light that can be used for things like printers

manufactured (man-yuh-FAK-churd) made something from raw materials

microchips (MYE-kroh-chips) tiny pieces of material that contain electronic circuits

nutritional (noo-TRISH-uh-nuhl) containing substances that are good for your health

static electricity (STAT-ik ih-lek-TRIS-ih-tee) electricity consisting of isolated motionless charges

technology (tek-NAH-luh-jee) use of science to solve problems

Index

air travel, 21
Apple, 11
architects, 12
artists, 22–23

bioprinting, 19
Body on a Chip, 18
buildings, 19–20

Canon, 9
Carlson, Chester, 4–6
cars, 12, 29
copiers
    color, 7, 9
    history of, 4–11

food, 24–25, 27, 29

Haloid Company, 6, 9
health care, 18, 19, 28, 29
Hewlett-Packard, 11
Hull, Charles, 9, 17, 28

IBM, 11
ink, 14

Kodama, Hideo, 10

laser printers, 9, 11

MakerBot, 13
Mercedes-Benz, 12

photocopies, 5, 6. *See also* copiers
printers
    3D. (*See* 3D printers)
    high-speed, 9
    history of, 4–11
    laser, 9, 11

Rapid Prototyping. (*See* 3D printers)

Shapeways, 16
space exploration, 21
Starkweather, Gary, 11
static electricity, 4–6

3D printers, 9
    advantages and disadvantages, 16
    early history, 10, 17, 28
    future of, 24–27
    at home, 24, 26
    how they can be used, 18–23
    how they work, 12–17
    timeline, 28–29
    vs. traditional printers, 12–14
    where to sell crafts made with, 16
    who uses them, 12

timeline, 28–29

Volvo, 12

Xerox Corporation, 6, 7, 9, 11
Xerox machine, 6, 8

Zazzle, 16